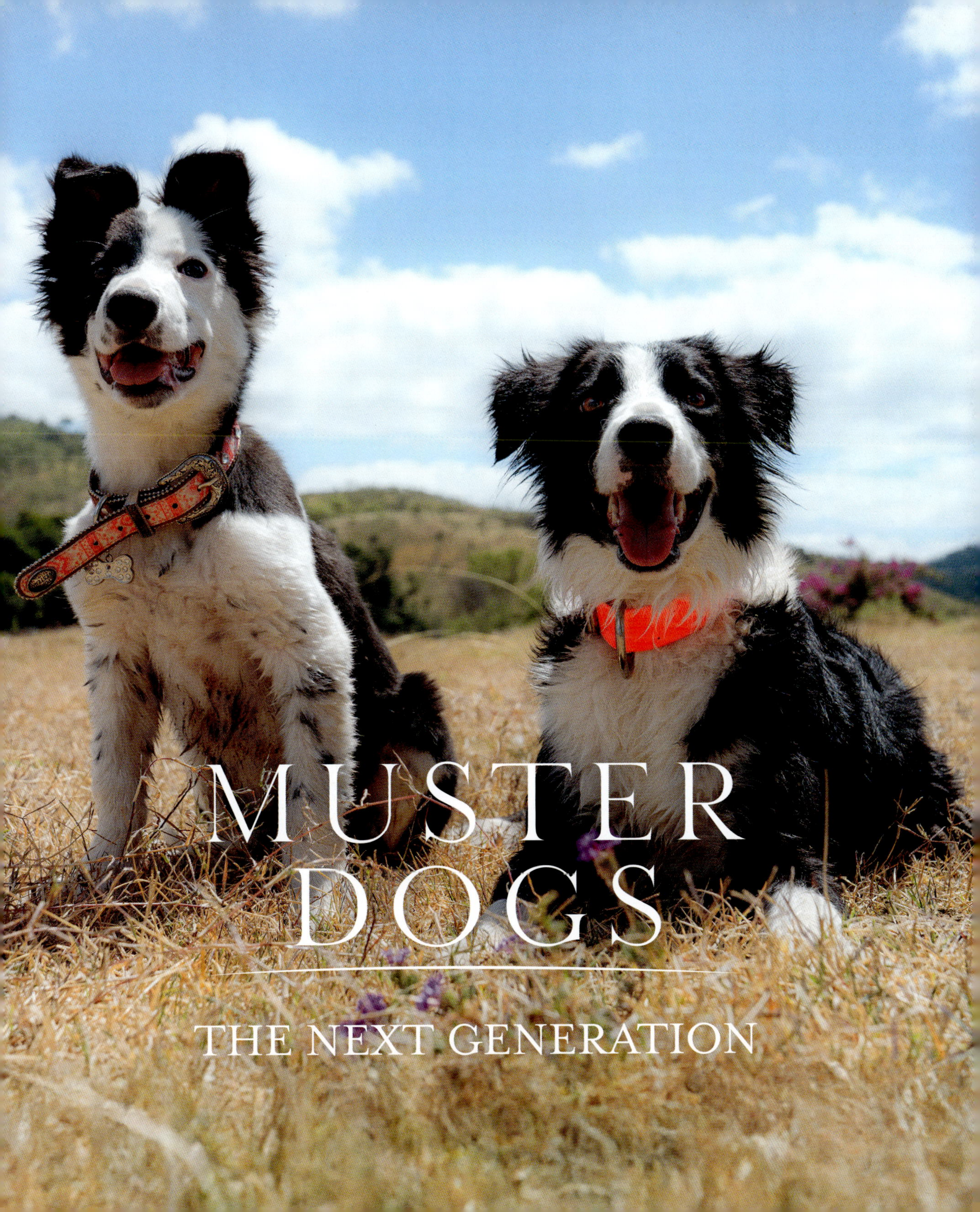
MUSTER
DOGS
THE NEXT GENERATION

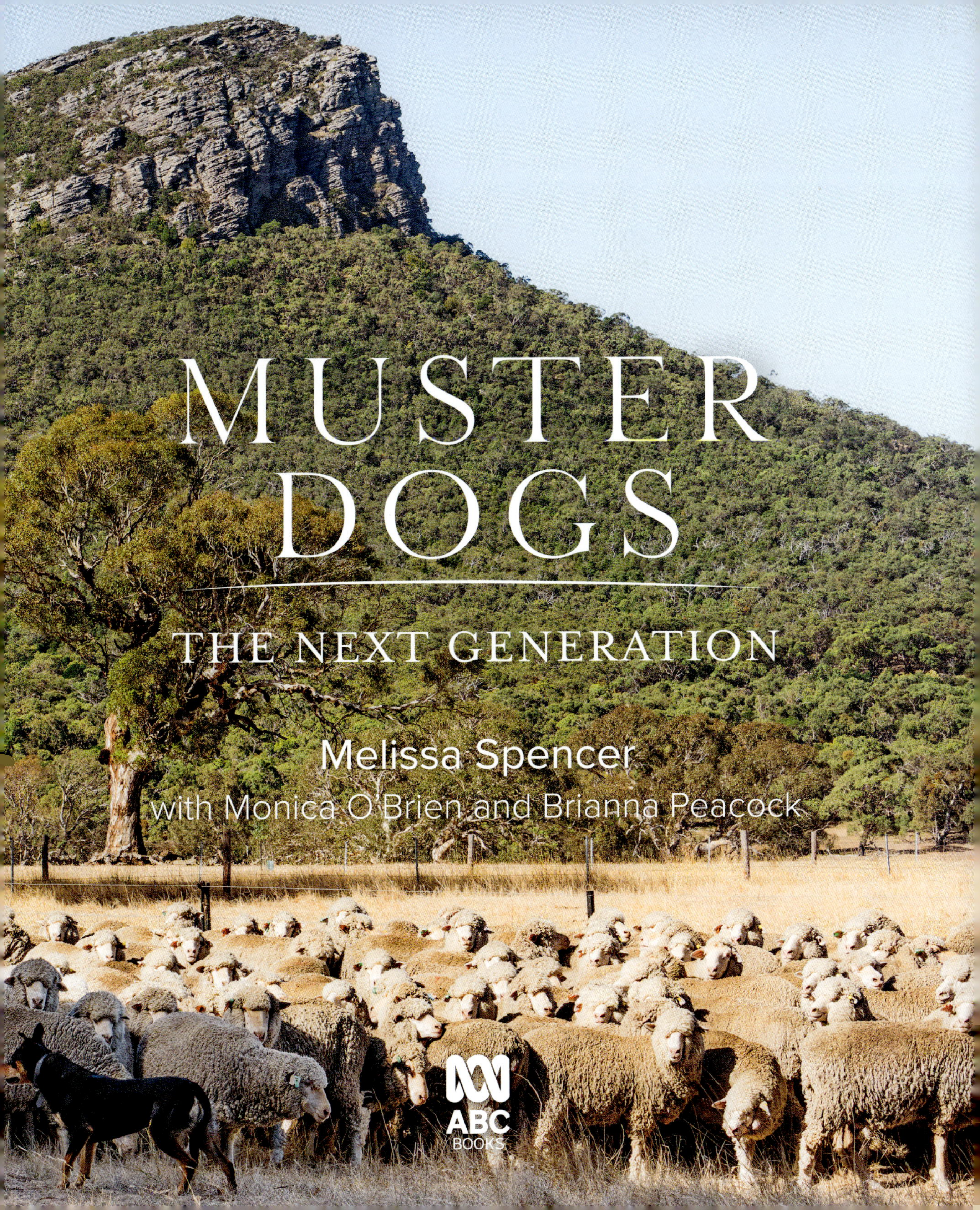
MUSTER
DOGS
THE NEXT GENERATION
Melissa Spencer
with Monica O'Brien and Brianna Peacock
ABC
BOOKS

CONTENTS

Foreword by Richard Glover

Has any other TV show supplied the pure pleasure of *Muster Dogs*? First you fall in love with the dogs; next you fall in love with the rural Australians who work beside those dogs. Then, whack, you realise you've become all misty-eyed about the country itself. This is Australia. These startlingly beautiful landscapes. These resilient, charming, hilarious humans, so often working in extreme environments.

And then those dogs. Those working dogs. So loyal. So smart. And, like most Australians, keen for a project through which they can show their eagerness, their talent, their joy at being alive and being here.

It's no surprise that reading about dogs can prove a teary experience. At the dawn of literature, a dog was already in attendance. In Homer's *The Odyssey*, the hero returns home wearing a disguise of a tatty beggar so that he might determine whether his household has remained loyal.

As Odysseus walks up to the gates of his palace, an ancient dog is lying on a dung heap outside the palace walls. It is the loyal dog Argos, trained as a puppy by Odysseus, but now abandoned – 'an object of revulsion, his master long since gone'. The dog alone sees through Odysseus's disguise. I've retold Homer's story before, but for dog lovers it's the story that expresses all we feel. In Daniel Mendelsohn's wonderful new translation:

When he sensed that Odysseus was close by
He wagged his tail and lay his ears down flat,
But no longer has the strength to come to his master.

Heartbreakingly, to protect his disguise, Odysseus cannot give the dog a passing nod. He just walks past – his torment expressed through a single tear. And then, says Homer:

Death's darkness then took hold of Argos, who
had seen Odysseus again, after 20 years.

Argos dies! And he dies because he knows his master is safely home. It's hard not to cry at the scene – I'm sobbing a little myself, now, as I type Homer's words, sentimental fool that I am. My own dog, Clancy, a kelpie who is often half-asleep on my bed as I sit at my computer, jumps up to investigate when he hears these strange choking noises.

Argos may be literature's best dog – although the loyal terrier in Tolstoy's *War and Peace* does come close. Oh, and I should mention Clancy who has written his own book. But now there are some new contenders. *Muster Dogs: The Next Generation* captures the joy of these glorious animals – animals with whom some of us are lucky enough to share our lives.

Clancy, by the way, is now back on my bed after the upset of me crying over Homer. He says, 'I agree. The dogs in this book are the best. Well, perhaps except for me.'

Foreword by Helen McDonald

From an uncharted beginning, *Muster Dogs* has captured the attention and hearts of Australians, especially children. The series has given kids an amazing insight into the challenges of training a high-quality working dog, the skills required to achieve success, and the rewards that result from commitment and dedication.

If *Muster Dogs* can encourage some of our younger generation into agriculture, this will help to turn the current tide in a more positive and profitable direction, and secure the production of high-quality food and fibre into the future. I find it extremely rewarding living on the land, working livestock with a capable team of working dogs that become your loyal companions. The benefits are far-reaching: you can work livestock singlehandedly, at a walk, creating a calm, co-operative environment; it also provides a better life for livestock, and knowing this gives me great pleasure.

I was fortunate enough, nine years ago, to purchase a female kelpie named Yipee Jess. She was our first 'tan head', traditionally called a saddleback. I mated her to Killili Pirate. The progeny of these two kelpies continues on, four generations later, with their admirable work traits, trainability and personalities suited to me and my husband Neil, plus so many others who have enjoyed working with this long line of kelpies.

Since the success of the first illustrated *Muster Dogs* book, *Mischief and Mateship*, photographer Melissa Spencer has dedicated her time to creating a second book, *Muster Dogs: The Next Generation*. She has captured amazing Australian landscapes and sunsets, and spent time in dusty sheep and cattle yards amongst happy, energetic puppies, patiently waiting for the perfect moment. The result is a book full of beautiful, captivating photos of kelpies and border collies with immense depth of detail.

As large areas of southern Australia continue to experience some of the driest conditions on record, our working dogs are a critical distraction of comfort and companionship, allowing us to keep our hearts open and our spirits strong as we tackle the challenges ahead.

INTRODUCTION

On farms and stations around this vast continent, it's not only skills being passed down, but a way of life. The bond between generations is forged through a combination of shared sunrises, hard yakka and the quiet rhythm of working side by side. Grandparents, parents and children are all part of this rich tapestry, each generation building on the contribution of the last, carrying forward the values, knowledge and pride that come with a life on the land. Their working dogs, too, carry the instincts, traits and temperament of their canine ancestors, with each new pup continuing the legacy.

Join us on a stunning photographic journey celebrating multiple generations – of our incredible working dogs, and of the families they love to grow, play and muster with.

Lilly

NEW PAWS, BRIGHT FUTURES

Each new generation of muster dogs comes with an abundance of promise and an element of surprise – you never quite know what you're going to get. What genetic lines will shape the origins of each new pup? And how will the humans that surround a pup influence its story?

As each pup grows, it will face hurdles along the way. Will it have what it takes to be a working dog? It's an exciting and often surprising time of discovery as these adorable creatures start their journey in the world.

For a grazier, finding a dog with the right attributes – such as being respectful on stock and having a good eye – is like finding a rare opal. It's also the chance to establish a new line, honouring the characteristics of the lineage that has gone before while adding their best dog's unique sparkle into the mix for generations to come.

PUPPY LOVE

The arrival of a new litter of puppies is a joyful occasion. As each new pup is born, it's accepted by its mother, nuzzled and nurtured. Some common personality traits will emerge in just a few weeks, then over the next 12 months attributes such as temperament, nature and working style will all take shape.

Birth. Warm and soft, siblings snuggle together for comfort and connect with the rhythms of each other's heartbeats.

5 Weeks. The pups are bursting with waggery, and the fluff-ball fun begins. Whether black, white or tan, short or long haired, these puppies are all on – or all off – as they rev through their gears. At top speed they roughhouse, gnawing at boots, sticks and ears, and in idle they flop and cuddle as they slip into a snooze.

10 Weeks. As they wrestle with gangling limbs, their individual personalities are emerging. A sense of confidence is growing and so is the desire to bond with a human who can take the pup on adventures.

A young pup's world is filled with wonderment and the pursuit of sensory satisfaction as it hounds smells, tastes and affection with ferocious tenacity. Handlers are hoping the pup's raw and instinctive desire to work stock can be refined and channelled into a well-mannered work ethic, allowing the duo to muster together with mutual respect, trust and enjoyment.

Only a few days old, these GoGetta working kelpie pups bred by Joe Spicer will one day grow into skilled, hardworking muster dogs.

THIS PAGE: Curled up in sleep, these tiny kelpies carry on the legacy of their dad, champion Muster Dog Banjo.

RIGHT: That smile says it all. This black-and-tan kelpie pup is ready to follow in legendary pawprints.

Cradled by Cilla, border collie Racquet, son of series 2 Muster Dog Ash Barky, is full of potential to follow in his mother's pawprints.

Muster Dog Blossom in relaxation mode.

Muster Dog Indi with talented daughter Summer by her side.

Fun in the sun! Carefree Sherwood kelpie pups play in between mealtimes.

These playful border collie pups, Clyde (left) and Marty (right), are as cheeky as they are clever. Testing their natural instincts, the pups show signs of greatness.

New pups, new possibilities. Jack's black-and-tan kelpies are bred with instinct and drive to work.

More than just a team, dog and grazier share a bond built on trust, respect and a whole lot of cuddles.

Clockwise from top left: Sam and Muster Dog Captain; Courtney and Muster Dog Blossom; Shaydie-Jane and Muster Dog Turner.

BLOODLINES AND BREEDING

Half the battle in breeding a good muster dog is picking the right genetics. The Australian working kelpie is bred for its hardiness and independent nature, suited to working sheep all day in the scorching heat under little to no command. On the other hand, the Australian working border collie is bred for its intelligence and willingness to take commands. Of course, there are variations within each breed, and many breeders select traits within their own lines that suit them best.

Mick Hudson, renowned dog trialler and breeder of MGH Australian working border collies, selects genetics that enable his dogs to shine in the sheep trialling world – they have compassion for their stock and are very biddable and eager to please. This is clear in our beautiful Muster Dog Indi, who has won dog trials with trainer Steve Elliott in Queensland.

Similarly, with Sherwood working kelpies, Helen and Neil McDonald select lines of kelpies that are biddable and personable, and with a lot of eye and style, as demonstrated by our lovely Muster Dog Pesto, who has beautiful natural work and is all style and class.

GoGetta working kelpies, bred by Joe Spicer, are dogs with spunk and heart – brave enough to work cattle on their own and go all day in the sheep yards, backing and barking. Our series 4 kelpie pups are perfect examples of the breed and will grow up to be the ideal companions and workers for their handlers.

Three generations of kelpies (left to right): Sherwood Marlin, Sherwood Reba, and now Muster Dog Banjo continues the working dog legacy.

Mother Marong Debbie (top right) and father MGH Snip (bottom right) looking across to their beautiful daughter, Muster Dog Indi (top left), who is a gorgeous combination of them both. From one generation to the next, the bond between border collies remains unbreakable.

LEFT: Marong Debbie and MGH Snip with daughters MGH Fairy and MGH Heidi.

TOP: Muster Dog Ash and her two border collie pups Racquet (left) and Boots (middle), beginning a new generation of herding brilliance.

RIGHT: Between mum Muster Dog Indi (left) and dad Barcoo Huey (right), pup Diamantina Steffie has big shoes to fill, but she's already proving she's up to the task.

LEFT: Full sisters MGH Heidi and MGH Fairy are opposites in every way, true yin and yang.

ABOVE: Proud mama Sherwood Reba surrounded by her three big, beautiful pups: Sherwood Cadbury (back), Sherwood Jean (front left), Muster Dog Banjo (right).

LEFT: GoGetta Betts is bred with class, style and eye, passing on these highly desirable working traits to his children, Muster Dog Banjo (opposite, bottom left) and Muster Dog Pesto (opposite, bottom right). GoGetta Dutchy (opposite, top) is their maternal grandmother.

Muster Dog Turner (middle) with half-sisters MGH Fairy and MGH Heidi.

The future of MGH Australian working border collies: these pups are ready to follow in their family's footsteps.

Working kelpies GoGetta Merck and GoGetta TV and (opposite) their two remarkable pups: Muster Dog Buruma (top), already showing unmatched instinct in the field, and Muster Dog Alfie (bottom), enjoying a well-earned moment of calm in the grass.

Brokenriver Bank, proud father to Annie, Chet, Luci, Gossip and Spice. A legacy of strength, intelligence and heart passed down to every pup.

Carefully bred for trialling and mustering, MGH border collies are shining examples of the Australian working border collie breed.

YOUNG DOGS, NEW TRICKS

For a grazier who has loved and trained their best dog and now has the opportunity to train a pup from that dog's bloodline, teaching the young dog new tricks is thrilling.

The importance of starting a pup early has universal consensus among our expert breeders and trainers. They all have different approaches to starting a pup, but agree on the basic foundations needed to train a muster pup.

Once their raw instinct kicks in, our muster pups are full of drive and desire to work. It's important to know how to direct the determination of a herding pup. It's a nuanced dance of pressure, response, then relief, ensuring the pup's natural instincts are honed and not over-dampened. This teaches the pup that if it listens to you, the reward is more opportunities to herd stock. Allowing the pup to grow as a confident muster dog requires short daily sessions working a quiet mob of dog-educated livestock, to ensure that the pup is learning in a controlled and safe environment.

But it's not all work and no play; spending quality time with your pup is just as important as training them. A strong bond will ensure that your dog will do anything you ask of it, come rain, hail or shine.

Joe Spicer with GoGetta kelpies Plan (above) and Muster Dog Alfie (right), teaching them the ropes of working with skill, instinct and heart. Every lesson is a step towards greatness.

Micky Moo

Training the next generation of working dogs. Neil McDonald teaches kelpie pup Ella the skills she'll carry for life.

Fearless from the start, kelpie pup Clag is nose to nose with the sheep, already showing the instinct and confidence of a true working dog.

Neil teaches kelpie Orange Blossom Special how to back sheep.

Muster Dog Captain is all smiles in trial champion and dog educator Mick Hudson's arms after his first day on stock.

Joe is impressed with up-and-coming kelpie Plan, who is excelling in his training sessions.

As the sun sets after a hard day's work, Mick and his Australian working border collies walk the paddock, showing what true teamwork looks like, no matter the hour.

At just four months, Muster Dog Indi's pups, Coffee (top) and Rocket (bottom left and right), are already showing the same talent and potential as their mum. Thanks to the guidance of award-winning dog trialler and trainer Steve Elliott, these pups are on the path to success.

Playtime is full of energy and excitement for these young pups. Muster Dog Blossom (this page); Muster Dogs Buruma and Turner (top right); Muster Dogs Captain, Buruma and Blossom (bottom right).

LEFT: Breeder Carolyn Hudson shares a special moment with her favourite dog, border collie Nikki.

RIGHT: Trainer Helen McDonald holding kelpie Mouse. Every moment of care and training shapes these dogs into incredible working kelpies.

YOUNG FARMERS, NEW WAYS

RB
COOLYNN KELPIES

Our food security is a global priority. Ensuring access to safe and nutritious food in order for populations around the world to grow and prosper is a huge challenge, and the burden of meeting this challenge falls largely on our primary food producers. Our Australian farmers and graziers work tirelessly at great personal sacrifice to feed an estimated 75 million people around the world every year. Working across weekdays and weekends, holidays and seasonal events, the responsibility to their animals, crops and communities is immense.

While the sunrises and sunsets are breathtaking, the all-weather days require resilience. Inspiring the next generation to work on the land and take on the responsibility and opportunity of feeding the world depends on making the work enjoyable, hopeful, sustainable and efficient. A grazier's loyal team of muster dogs is critical in achieving these goals, and together they can face the joys and challenges each new day brings.

A FAMILY AFFAIR

It's all hands on deck at these family farms, and for many young kids there's no better place to grow up. Some families have been working with dogs for generations, while others have only just started, but they all have one thing in common: kids who are keen and capable. With their handy dogs by their side, some budding young graziers can move mobs of livestock safely on their own, a feat that many adults would struggle to achieve.

The beauty of having quiet, cooperative, dog-educated livestock is that they are safe for children to be around, and this provides a dynamic and engaging learning environment. Working with dogs and livestock from a young age gives children a sense of accomplishment and ensures the future of the industry is in safe hands.

There is nothing more moving than watching a child work their dogs alongside a parent or grandparent, knowing they're walking in their relative's footsteps, and are already leaving a few of their own.

LEFT: The next generation of Muster Dogs and muster kids, growing up side by side and learning together. (Left to right): Greta, border collie Racquet, Annie, Cilla, border collie Rosie, Sidney, kelpie Pip and Muster Dog Ash (front).

Young Flynn is training his own dog and keen to help his mum, Kim, get the lambs back to the yards for processing.

The next generation of farmers, already showing love and care for the pups that will one day work alongside them. Mara (left) has her loose-lead walk with Muster Dog Pesto down pat. Sidney and kelpie Pip (right) and Greta and her border collie Boots (above) are naturals at puppy wrangling, calmly getting their pups to sit for a photo.

Frank's son Scott and grandson Hudson working young kelpie pup Bono and old kelpie Barney through the quarter bubble.

Smiles, dogs and the next generation. Jack with his daughter Mara (top). Marlene's son Daniel and granddaughter Amelia (bottom).

LEFT: At just 10 years old, Frank's grandson Cody is already displaying fantastic skill, working both cattle and dogs with precision like a seasoned pro. The future is in great hands.

RB
COOLYNN KELPIES

ABOVE: Marlene and granddaughter Amelia with Muster Dog Hudson, capturing a moment of love, legacy and family, all wrapped up in one perfect shot.

LEFT: Flynn bonding with kelpie Duke.

THE CHANGING FACE OF MUSTERING

Once, the image of a muster on the horizon featured a sea of riders on horseback, station hands spread wide across dusty paddocks. These days, it's more likely to be a single grazier in a buggy or on a motorbike with a team of loyal dogs by their side, skilfully bringing in a mob with precision and grace.

The power and value of a team of well-bred muster dogs is recognised across the nation as graziers experience the joy and profitability of calm, dog-educated livestock. Herding dogs may have been around for hundreds of years, but the way they are worked and used is changing. The next generation of graziers is embracing this change, and with fewer hands to call on, they are turning to their dogs more than ever. Whether they're moving cattle daily in a cell-grazing system, mustering the ewes in for drenching, or educating fresh weaners, a team of muster dogs is more than just a tool, they're teammates, teachers and, in many ways, the heart and soul of the operation.

While technology has advanced and changed the way farms operate, the bond between a handler and their dogs remains timeless. In this ever-evolving landscape, the muster dog is a constant, a living link between generations past and the future of the land.

The late afternoon shift: muster dogs, Renee and her family working together to move the mob.

Another successful day in the sheep yards for Jack and his kelpies – heading home for smoko with Muster Dog Pesto and kelpie Sushi.

Eighteen-year-old Nathan (left) shows incredible wisdom and a passion for the land. He leads the way (above) with his mum, dad and loyal Muster Dog Chief by his side. Handsome Muster Dog Chief (right) is never short of a cheeky grin.

James and Dean take the dogs home after a big day.

Kelpie Moss and Muster Dog Pockets sharing a special moment. The connection between Renee (above and top left) and her working dogs is precious.

Newcomer Shaydie-Jane with her special Muster Dog pup Turner (left). Kicking goals and holding her own while working with kelpie Dutchy (above).

ABOVE: Blythe gives Muster Dog Banksi the biggest hug in the season 3 finale, so proud of how far he's come and everything they've accomplished side by side.

RIGHT: Captured in a quiet moment, Zoë and her kelpie Tango share a heartwarming bond. Any time is a good time for a pat.

Against the breathtaking backdrop of Bothwell, Tasmania, Russ and his family watch proudly as kelpie Ted gets to work.

QUITE THE SCENE

There's something deeply serene and striking about the vast open spaces and varied terrain that so many Australians call home. From lush green mountain ranges to the red dust swirling in the outback, these stunning scenes reveal a shared rhythm, humans and dogs moving together with quiet understanding. They paint a vivid picture of a nation grounded in the hard work and enduring partnership between its people and their loyal four-legged mates. And there's always time for play when the day is done.

RIGHT: Muster Dog Pockets, mid-leap, as she jumps straight into Renee's arms. Pure joy, trust and a bond that's unshakable.

Shall we dance? Muster Dog Roxy and Max show off their fancy footwork before heading to the shore.

TOP: Renee's devoted pack, all eyes on her, ready and grinning from ear to ear.

RIGHT: When work is done, it's time to jump for joy. Muster Dog Chief showing off his playful side.

With this stunning backdrop, Muster Dog Hudson (left), kelpie Chewie (right), and Marlene on horseback work in perfect harmony. Everything comes together for a smooth, seamless muster.

LEARNING FROM THE OLD GUARD

The opportunity to learn from the old guard is a gift valued by regional families around the country. With decades of experience under their belts, older farmers and graziers have a wealth of knowledge, skill and insight to share, from reading weather patterns and handling stock, to managing country with care and instinct. From one generation to the next, this knowledge is passed down through lessons in the yards, stories over the ute bonnet, and time spent working side by side.

While the older generation offers time-honed wisdom, the next generation is forging its own path, blending tradition with innovation and often teaching the old hands a thing or two along the way. Together, with their trusty working dogs, they're shaping a smarter, stronger future for life on the land.

THE GETTING OF WISDOM

Generational wisdom is about more than just practical skills; it's a deep understanding of land, life and legacy. For our young farmers, attaining this knowledge is a rite of passage in which the combined lived experiences, traditions and lessons learned over time are passed down from elders as a source of strength and guidance.

RIGHT: Drawing on a lifetime of knowledge and experience, Frank instructs 18-year-old Nathan. Mentorship, tradition and the future of mustering all in one powerful moment.

A picture of success: Kim and Muster Dog Banjo with Helen and Neil, the mentors behind their championship journey. Where skill, passion and guidance meet to create champions.

Combining drone power with the valuable knowledge shared by fellow graziers. A perfect blend of innovation and tradition.

Father and daughter Tom and Shaydie-Jane share more than just a moment. Passing down knowledge, life lessons and a deep connection to the land.

Muster Dog Hudson's cheeky grin is all you need to brighten the day.

Marlene shares her wealth of knowledge and the lessons of the land with her son Daniel. A legacy that will continue for generations.

Muster Dog Hudson is focused and ready, waiting for the call to action from Marlene.

James, Renee, Dean and Greg show off their stock skills as they yard the mob for drafting.

Dad Greg and daughter Renee with their dogs. The perfect mustering team.

Kelpie dogs Mango and Buffer are all smiles, showing their love and loyalty to Greg.

Nathan yarding cattle with his dogs, kelpie Indi and Muster Dog Chief, and the helping hand of his grandfather Alan.

MUSTER MENTORS

Watching a young dog work alongside an older dog is like seeing instinct come to life. The pup mirrors every stride, movement and turn, learning from the best in a silent exchange. With seasons of experience behind them, these old dogs carry a calm authority and quiet confidence that can't be taught, only absorbed. Whether they're stepping in to guide a mob or simply sitting off to the side, an older dog's presence is a steadying force. Even in their twilight years, working dogs are loyal and generous with their wisdom, passing it on with every movement. For the young dogs at their side, they are the best teachers a pup could ask for.

RIGHT: Young Muster Dog Pesto strides in sync with seasoned kelpie Lola.

TOP: Seasoned kelpies Dusty and Barney have earned their rest, but are always ready for the next muster.

RIGHT: Border collie Freedom keenly works stock, with experienced kelpie Marlin ready to back him up.

LEFT: Muster Dog Ash works alongside seasoned border collie Molly, picking up the skills and soaking in the wisdom of the more experienced dog.

Muster Dog Pockets and 11-year-old kelpie Tap. They might be generations apart but they share the same drive and desire to work.

Mother and daughter kelpies Lola and Tofu working together.

No matter their age or the rough landscape, working border collies are always ready for the next job – nothing will slow them down.

With a focused gaze, these three kelpies, Nelson, Pumpkin and Flex, muster with a steady yet commanding demeanour, showing compassion for their stock. Their calm presence creates a good environment for a dog to learn in.

Muster Dog Chief and kelpie Trapper
patiently wait for the next job.

LIFE ON THE LAND

Living out in the bush is not for the fainthearted, but ask any grazier and they'll tell you there's nothing better. Life on the land has its challenges; mother nature is not always kind. Yet despite the hardships, the tranquillity and freedom of country life offers rich rewards. There's joy in watching calves grow fat and feisty alongside their mothers on the green grass, in the smell of the first rain hitting the dust after a dry spell, and in the golden glow of a sunset over dusty stockyards after a long day. With a couple of good working dogs by your side and the wide open country all around you, it's truly a magical life, one that's hard to beat.

A TIMELESS PARTNERSHIP

For centuries, working dogs have toiled alongside humans to move livestock and manage the land, forming an essential partnership. The relationship between a grazier, their dog and the land is built on trust, instinct and countless hours spent side by side. Whether it's a well-timed whistle or a dog's steady eye on a mob, there's a synergy to this relationship that never fades. In a world that's always changing, this collaboration remains – timeless, dependable and at the heart of life on the land.

Kelpie Bank stylishly working the sheep as the sun sets. These dogs and their handlers make every moment look effortless and beautiful.

Leaping into action, these kelpies are always ready to jump into their work, whether it's herding stock or hopping into the buggy.

Brothers Muster Dog Hudson (left) and Muster Dog Chief (above) showing their similarities as they work stock.

A seamless flow of sheep, expertly directed by sharp eyes and steady paws. This is working dog magic!

She may be goofy but Muster Dog Pockets has serious skill when it comes to getting the job done.

Dog-educated cattle make for a smooth move.

A QUICK DIP

For a muster dog, water is the ultimate source of pleasure. Whether it's jumping in the creek after a stick, or having a quick 'bogey' in the dam on a hot day while working in the paddock, there's no better way to cool off. After long hours on the job, it's a chance to refresh, recharge, shake off the dust and get ready for what's next.

Border collie Sweep and Muster Dog Chief take a quick bogey in the trough.

Time for the classic shake-and-soak-the-humans routine.

Living the dream: mud, water and a well-earned rest. (Left to right): Border collies Gus, Tri and Foss.

A GLIMPSE INTO THE FUTURE

Those who came before, both graziers and their dogs, have laid a strong foundation, passing down valuable skills and traditions that have set the future of mustering on a promising path. While no one can predict exactly what lies ahead, there's hope and confidence that the next generation will carry the legacy forward with pride. Advances in technology and new methods will continue to change the industry, but one thing remains certain: no grazier will ever be alone, not with their loyal dogs by their side, at the ready to get the job done.

LEFT: With the setting sun behind them, Joe and his kelpies finish the day's work and head home.

ACKNOWLEDGEMENTS

Creating *Muster Dogs: The Next Generation* has been a labour of love, and it would not have been possible without the support, guidance and contributions of many wonderful people.

First and foremost, we would like to acknowledge the Traditional Owners of the Lands on which *Muster Dogs* was filmed and these images were taken.

Thank you to the participants, experts, graziers and their wonderful families who have opened their homes and hearts to us for this book and over the many years of filming *Muster Dogs*. Your stories and dedication have been the backbone of this series.

We would also like to express our gratitude and pay tribute to the agriculture industry and the tireless efforts of families, both new and old, along with their remarkable working dogs.

In particular, we would like to express our gratitude to the breeders of these wonderful working dogs. Special thanks to Joe Spicer for breeding the

GoGetta working kelpies, Mick and Carolyn Hudson for breeding MGH Australian working border collies, and Neil and Helen McDonald for breeding the Sherwood working kelpies featured in *Muster Dogs* and this book.

A special thanks to our incredible crew and all those involved in the creation of the series and this book.

Thank you from the Ambience team to the talented Melissa Spencer, whose work is featured in this book. Your skill and passion have truly captured the essence of *Muster Dogs* and the next generation of working dogs, graziers and farming families.

Finally, to the readers, thank you for taking the time to explore these pages. We hope they leave you with a sense of wonder and pride for the future, seen in the eyes of each young muster dog, in the hands of the graziers and families shaping tomorrow's land and livestock, and in the enduring spirit of Australian agriculture.

LOCATIONS

Bingara, New South Wales: pp. 12–13, 18–19, 112–113, 128–129, 186

Bothwell, Tasmania: pp. 104–105

Central Highlands, Queensland: pp. 4–5, 74–75, 92–93, 138–139, 168–169

Clermont, Queensland: pp. 84–85, 114–115

Dubbo, New South Wales: pp. 6–7, 14, 66–67

Dunkeld, Victoria: pp. 2–3, 146–147, 164–165, 180–181, 184–185

Glenthompson, Victoria: pp. 152–153

Kaniva, Victoria: pp. 80–81

Kingaroy, Queensland: pp. 132–133, 182–183

Sherwood, South Australia: pp. 120–121, 142–143, 160–161

PHOTO CREDITS

All photographs © Stock Chick Films | Melissa Spencer, except for the following:

Olivia Rousset: p. 8

Brook Rushton: p. 102

Trudy Obst: pp. 119, 123 (top right)

ABOUT THE AUTHORS

Melissa Spencer

Melissa Spencer (Stock Chick Films) is a videographer/photographer and grazier from Capella, Queensland. With a working dog team of her own and years of experience working on cattle properties in Central Queensland, she has a deep passion for showcasing rural stories and the beauty and skill of Australia's working dogs through her lens. As a visual content creator specialising in video and photography of rural life and culture, she has also worked as a camera operator and photographer on the TV series *Muster Dogs*.

Monica O'Brien

Monica is an experienced creator and show runner across many genres including Factual Entertainment, Drama and Children's Entertainment. Over the past 17 years Monica has worked with Ambience Entertainment in production and development across a broad slate of programming, including *Magical Tales*, *Drop Dead Weird*, *Muster Dogs*, *Barrumbi Kids* and *The Garden Hustle*.

During her time writing, directing and producing the *Muster Dogs* series, Monica has travelled Australia meeting incredible people, learning about their communities and filming their beautiful dogs. Through that process she has been enriched by the stories of working dogs and the lives they live.

Brianna Peacock

Brianna is an experienced marketing manager with Ambience Entertainment, who has developed a deep respect for the working dogs, graziers and locations of regional Australia. She has contributed to bringing their stories to a national and global public through her work on the *Muster Dogs* TV series and the hit show's companion illustrated books.

HarperCollins*Publishers*
Australia • Brazil • Canada • France • Germany • Holland • India
Italy • Japan • Mexico • New Zealand • Poland • Spain • Sweden
Switzerland • United Kingdom • United States of America

HarperCollins acknowledges the Traditional Custodians of the lands upon which we live and work, and pays respect to Elders past and present.

First published on Gadigal Country in Australia in 2025
by HarperCollins*Publishers* Australia Pty Limited
ABN 36 009 913 517
harpercollins.com.au

HarperCollins*Publishers*
Macken House, 39/40 Mayor Street Upper
Dublin 1, D01 C9W8, Ireland

A catalogue record for this book is available from the National Library of Australia.

ISBN 978 0 7333 4426 8

Edited by Katherine Hassett
Cover and internal design by Jude Rowe, Agave Creative Group
Cover photographs © Stock Chick Films | Melissa Spencer
Colour reproduction by Splitting Image Colour Studio, Wantirna, Vic
Printed and bound in China by 1010 Printing on 128gsm matt art

6 5 4 3 2 1 25 26 27 28 29